# Safety and Health Savvy

In an ever-changing world, understanding the essentials of safety and risk management is more crucial than ever. This concise guide not only revisits fundamental principles but also integrates the latest developments in wellbeing and mental health. Designed to be as user-friendly as possible, it simplifies complex concepts into practical advice that can be easily applied in day-to-day life. This book offers valuable insights for anyone looking to enhance their understanding of personal safety and wellbeing in the workplace. Readers will find a fascinating combination of key safety theories, grounded in the latest safety thinking, alongside actionable tips and methodologies.

Each chapter is designed to be accessible, featuring everyday case studies from a variety of industries that bring the concepts to life. From personal risk management to insights based on lived experience from top experts in health and safety, this book encourages readers to not only practice these principles but to explore further on their own, thus giving them a rounded and contemporary view of how health and safety can be considered in today's modern workplace.

*Safety and Health Savvy: A User's Guide to Thriving in the Workplace* is perfect for professionals at any career level in occupational health and safety, human resources, and business and management.

# Safety and Health Savvy
## A User's Guide to Thriving in the Workplace

**Tim Marsh**

CRC Press
Taylor & Francis Group
Boca Raton London New York

CRC Press is an imprint of the
Taylor & Francis Group, an **informa** business

Designed cover image: Shutterstuck

First edition published 2026
by CRC Press
2385 NW Executive Center Drive, Suite 320, Boca Raton FL 33431

and by CRC Press
4 Park Square, Milton Park, Abingdon, Oxon, OX14 4RN

*CRC Press is an imprint of Taylor & Francis Group, LLC*

© 2026 Tim Marsh

ISBN: 978-1-032-97603-7 (hbk)
ISBN: 978-1-032-96045-6 (pbk)
ISBN: 978-1-003-59442-0 (ebk)

DOI: 10.1201/9781003594420

Typeset in Times
by Deanta Global Publishing Services, Chennai, India

I'd like to quote the late great Roy Castle (a famous UK entertainer and TV host) ... who sang a song suggesting that *dedication* is the thing we most need to thrive and succeed ... Please see Chapter 1 (and Roy, setting the tone of this book perfectly, on YouTube!) for more details ...

And for my lovely friend Nick, with whom I walked and talked and tried to make sense of the world.

# Contents

# Introduction

I've been working in Health and Safety for more than 30 years now and as I write in 2025 – and please forgive the boast – I still get lots of compliments about my levels of passion and commitment.

So, here's a bitterly ironic story to start a book with as it helps explain what has always energised me. (A motivation and value articulated perfectly at the end of this introduction by a lawyer of all people!)

Back in the UK in 1972 a Lord Robens published a report that was the basis of the 1974 Health and Safety at Work Act which suggested that the onus of risk management be in the hands of the risk owners. Consequently, although the UK is average at best in a number of related and important fields (wellbeing and mental health especially) we're actually a world leader when it comes to safety.

Consequently, to some, Robens is a huge safety hero ... but *not to me* however.

You see, while in charge of the National Coal Board, he presided over the Aberfan disaster of 21 October 1966 where an enormous waste tip at the Merthyr Vale Colliery became slurry after heavy rain and slid down the mountain burying Pantglas Junior School and several houses. Some 144 people were killed. 116 of them children aged just 7 to 10. 28 adults also died.

(It's personal to me as my mother was dropping me at a very similar school on that same morning about 15 miles away.)

In truth, Lord Robens would have been as oblivious as we were to the events about to take place. In those days it was totally acceptable to think that no news was indeed good news so it would be wrong to retrospectively blame the National Coal Board's chairman for a culture where proactive questions like 'what's the worst that could happen?' weren't asked.

It was genuinely 'another time'. However, ...

Robens didn't rush straight to South Wales on news of the tragedy but delayed setting off for 24 hours so that he could continue with his planned schedule and, elegantly berobed, be installed as Chancellor of the University of Surrey. (Well, first things first ....)

When he did eventually arrive in Wales, he promptly announced that he could say with *certainty* that the tragedy had nothing to do with the National Coal Board (NCB). (At this point he hadn't even been briefed and therefore had, frankly, absolutely no idea whatsoever what had caused it.) Later he insisted that it was impossible to have known that there was a spring in the heart of the tip – and this remained the NCB's official view throughout the subsequent 76-day tribunal of inquiry.

But *everybody* knew that there had been tipping over not one but two springs for *years* and both were actually on any Ordnance Survey map. Robens eventually conceded this point at the very end of the Inquiry. (It was angrily pointed out that if he'd done that on day one huge amounts of cost and upset would have been avoided.)

It was only on day 70 of the Inquiry that Robens finally admitted under cross-questioning that 'yes', he had to concede that the event was 'reasonably foreseeable' and that therefore the National Coal Board was culpable. It is reported he appeared utterly affronted to even be challenged and was called a self-important, incompetent, unhelpful and contradictory witness.

He 'offered'* to resign but kept his job and indeed, *nobody at all* was prosecuted, dismissed or even demoted as a consequence of the disaster (*this 'offer' to resign submitted by him along with suggested reasons why it *shouldn't* be accepted. Classy Alf).

The NCB then took money from donations to help pay for the clear up operations and the making safe of other dangerous tips. They even had the now homeless families pay towards the temporary accommodation that was set up.

Perhaps worst of all the NCB planned to have parents testify as to how close to their children they were when determining compensation … (yes, *really* and you're swearing now, aren't you?).

On all these issues the villagers fought back hard and social historians consider their refusal to be cowed a major event of social history. Deference to authority was crumbling globally and the times were indeed 'a changing' to quote the Bob Dylan song released a few years earlier.

**Postscript.** A little later in Lord Robens' career, drawing on these events as he must surely have done, he wrote a really excellent report. Although improved workplace safety was also an element of the *general* shift in attitude who knows how many lives have been saved, or not ruined, as a direct consequence of the report's impact over the decades?

Ironically, it's almost certainly many more than were lost on that terrible October morning.

You may think this a quite angry way to start a light-hearted book so please consider the closing words from Desmond Ackner, quoting George Bernard Shaw, council for the families at the tribunal:

> *The worst sin toward others is not to hate them, but to be indifferent to them: that's the essence of inhumanity.*

This book strives hard to describe in simple terms the practical steps we can take to **not be indifferent** to our colleagues or, indeed, to ourselves. Not to take anything for granted. Not to accept things as they are and 'have always been'. Not to accept 'because I say so'. And especially not to accept that 'fate' is all powerful – because, even today, lots of people will pop up on the TV suggesting that 'accidents just happen' and that the 'elf and safety gestapo' have gone mad. Can I suggest two things:

One: 'actually, no accidents don't just f***ing happen'. There's barely an accident in history that couldn't have been easily prevented – often in any number of ways. The trick is to spot and action these solutions pro-actively *before* the event – not identify the easy ways it *could* have been avoided in hindsight.

Two: sometimes the red tape is indeed unnecessary but nine times out of ten a reference to 'elf and safety madness', translates as 'I just want to crack on here, make my money and can't be assed with all this safety stuff … there's a "bigger picture" to think about and anyway we'll most probably be OK … again'.

But sometimes it's not OK and sometimes that'll be life changing for a person and their family. Sometimes for an entire village or even a town (see, of course, tragedies such as Bophal).

Because have you ever noticed that people pointing at 'the bigger picture' basically mean, more often than not, making money or avoiding hard and/or expensive work even when that'll ensure other people are safe?

That sounds pretty 'us and them' sceptical I know but the good news is that safety and health just isn't an either/or issue. It really is a 'win–win'. There's nothing in this book that isn't good for the bottom line as well as the workforce.

There's nothing in this book that isn't good for the bottom line as well as the workforce.

It really is a 'win–win' that includes business success and sustainability.

(And there's a very good chance you're holding a copy because your employers agree with that!)

However, it very explicitly aims at the people who do the hard, risky work so as to maximise their chance of going home in one piece, uncontaminated – and with a big smile on their face as they relish a good day at work.

Because the world is a very dangerous place: Though in the UK less than 200 are killed at work annually in a country of 67 million is pretty impressive as a head-line – other figures are more sobering. Around 2,500 will die on the roads, 3,000 in accidents in their own homes and 12,000 because of exposure at work – typically many years previously (and that later figure thought a *conservative* estimate).

Then there's the 4,200 or so working age people who will take their own lives.

And that's not to even mention the 'life changing' injuries and illnesses. This phrase means the loss of an eye or eyes, loss of limbs, paralysis, brain damage or such as heart and/or lung problems which makes even climbing stairs impossible – let alone a kick around with the grandkids.

So, this book isn't about being scared of all risk and never leaving your house … it's about having clever personal risk management habits so you can 'not just survive' … but ideally *thrive* with 'passion, compassion, humour and style' as Mia Angelou put it.

Nothing can guarantee good luck and happiness sadly. But! I'll make the case that we can have far more control of how much luck we might need – and how likely we are to be happy. We can load the dice in our favour.

You just need to be a little Health and Safety Savvy …

Enjoy!

'*It's not about surviving, it's about thriving with passion, compassion, humour and style*'.
(*Maya Angelou*)

# About the Author

**Tim Marsh** is a is a Consultant, Chartered Psychologist and a Chartered Fellow of the Institute of Occupational Safety and Health (IOSH). Widely considered a world authority on the subject of behavioural safety, safety leadership and organisational culture, Tim has worked with hundreds of companies around the world, including the BBC, the National Theatre and the European Space Agency, as well as the usual list of blue-chip organizations from manufacturing, utilities, food and drink, oil and gas, and pharmaceutics. He has keynoted and chaired dozens of conferences around the world including the closing keynote at the Campbell Institutes inaugural International Thoughts Leaders event in 2014 and was made visiting professor at Plymouth University, UK, in 2015.

# 1 The Nature of Luck

The golfer Gary Player is famous for the quip 'it's funny, the harder I practise the luckier I seem to get' when people observed he often got a 'lucky' bounce. I always pictured him practising his bunker shots at dusk until his hands bled – but it turns out he was a real gym bunny decades before Tiger Woods and other golfers followed suite.

Even more than practising, the effort and sweat involved in lifting weights is an excellent metaphor for what this chapter is about: that, on balance, you get the luck you tend to deserve depending on the graft you put in. (See also Roy Castle!)

Sadly, we none of us get any guarantees either way … but you *can* load the dice in your favour …

Let's start with a consideration of the incident that fills hospitals – the simple fall. What happens when you fall over? Well usually nothing at all except a little embarrassment if someone sees us. Less often we'll suffer a minor injury like sprain a wrist or scuff a knee. Less often again, we'll hurt ourselves properly – maybe a broken wrist or elbow or we'll hit our head and concuss ourselves, And less often *again* we'll hit our head and suffer terrible brain damage and/or die. (Michael Schumaker and Natasha Richardson would be, at the time of their accidents, young and healthy examples of the later.)

Whatever we're talking about the data will look a little like the diagram below. (Do not worry about the exact ratio – that'll of course change depending on the situation and behaviours in question. It's the *principle* that you need to remember: that *on average* for every serious incident there will be dozens/ hundreds of near misses and hundreds/ thousands of unsafe acts.)

So, you may *personally* be unlucky quickly – or never at all.

The thing is, who exactly will be unlucky is impossible to predict but how many people are going to get unlucky and who is *most likely* to get unlucky I *can* tell you!).

DOI: 10.1201/9781003594420-1

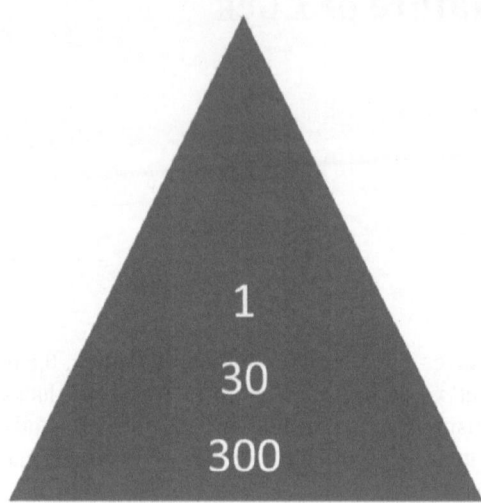

Here's a simple example from the world of work. The likelihood of falling down the stairs on an oil rig is about 100,000 to 1 if you're not holding the handrail. (But that would be a bad thing as the stairs are both steep and metal.)

Two things flow from this data – the first that a lucky person could work an entire career off-shore, never holding the handrail and never falling. The second is that if a supervisor wants an excuse to turn a blind eye to this unsafe act, then phrasing it as 'ah come on, it's *only* a 100,000 to 1 shot' is all that's needed.

But here's the thing: the stairs on a typical oil platform are used about a million times a year – so if no one holds the handrail we're looking at 10 falls give or take. If, however, 90% hold the hand rail, a fall a year give or take. But if 99% hold the hand rail we're looking at a fall every 10 years or so. It's a really simple numbers game.

And here's a second thing you might not know that illustrates the truth of this: more people have been killed in falls in the UK off-shore industry than in all the process safety issues combined (Piper Alpha included, even though some 167 died that single night.)

*Simply holding the handrail is a __very__ important safety behaviour.*

And, related, it's the number one cause of fatalities in the home too. For example, last year some 2,000,000 plus people in the USA hospitalised themselves by failing to hold the handrail/falling down stairs.

### Gravity

Indeed, **any** bad habit where gravity can come into play is a big issue. As above, climbing on or up anything at all unsafely fills the hospitals. From roofs to ladders to horses to chairs when hanging Christmas decorations. (So add another 500,000 or so from the UK to the figure above for the USA as our population about one-quarter the size.)

An exercise ... quickly name a handful of famous people who made the papers for having an accident:

..............................................................................................................
..............................................................................................................
..............................................................................................................
..............................................................................................................
..............................................................................................................
..............................................................................................................
..............................................................................................................
..............................................................................................................
........................................................................

How many of them will have had a bad interaction with gravity? Indeed, did *any* you named not?

I'm thinking Ozzy Osbourne (who fell off a bike), Christopher Reeve (who fell off a horse), Eric Clapton's son (who fell from a window) – and, as above, Michael Schumaker – who survived his entire Formula One career unscathed but then fell off his skis.

So, some top 'gravity' related tips:

- Hold the handrail.
- Do not stand under that.
- Wear a hard hat even when it only *might* be needed.
- Always step down squarely.
- Do not stand/climb on that*.

*And if you simply can't resist "standing/climbing on that" (see the following chapter on temptation) then please apply the golden rule that you always, *always* risk assess where you might fall and what you might fall onto.

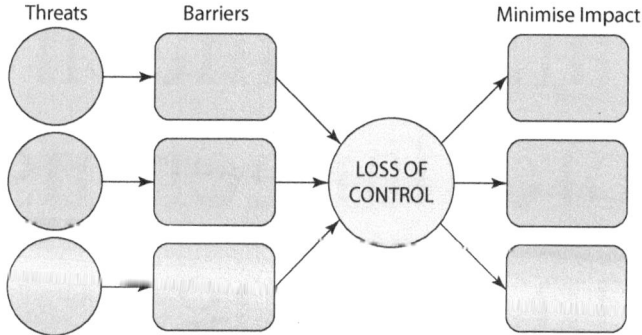

This is the right-hand side of the classic bow tie approach to risk management. A bow tie analysis tries to predict what might happen. On the left hand-side we analyse the reasons why it might – so we can try and stop it at source. On the right-hand side a consideration of how we might minimise its impact if it does happen. Speed limits and cameras and driving tests on the left and air bags, crumple zones, anti-lock-breaking and lane barriers on the right would be a simple driving illustration.

Thinking in advance if it goes wrong – as sooner or later it probably will so 'when' better – how can we minimise the hit? (The left-hand side is of course full of items like 'don't climb on unsafe things' or 'don't drive too close to the car in front'.)

A very simple ***domestic*** example:

*if you're changing a bulb in a bathroom on a rickety chair – or some such – and might well land on the toilet bowl if it gave way or you slip ... then at least be certain to put the lid down!*

# Here's a Question

Why is riding a motorbike statistically so much more dangerous than driving a car?

Answer: 'because you can't fall off a car'.

**Same Principle – Non-Gravity Related**

Here's some other everyday non gravity related behaviours where trusting to luck is a bad idea:

- Driving too close to the car in front.
- Driving too fast for the conditions.
- Driving distracted/angry/tired or when rushing.
- Not wearing Personal Protective Equipment (PPE) correctly.
- Not making eye contact with lorry/plant drivers whilst crossing a yard.
- Not wearing PPE when signs suggest you need to (see later chapter as this very, very important indeed).

And ...

- Not being reasonable and respectful to your boss/colleague/report/client/customer/partner etc. etc. ... because every action has an equal and opposite reaction ... we'll get back to this!

This isn't even close to a comprehensive list of course but you get the idea, I'm sure. It's all about loading the dice.

Or you can do what the hell you like and trust to luck. After all, even smoking only has a 50–50 chance of killing you. Optimists can focus on that 'only' – perhaps modelling themselves striding fearlessly from the helicopter (like in the famous film) announcing:

'I love the smell of napalm in the morning'.

Whilst all around people cower in fear. It was certainly a pretty cool scene ... but he was an actor in a film. In a real-life battle, Nelson's last words before being shot by a French sniper were:

'Don't worry men they can't hit anything from that distance!'

**We Need Fleetwood Mac!**

It may be an urban myth, but you may have heard that the Fleetwood Mac song 'The Chain' is banned from the radio. Famously used as the Formula One theme music it was suggested drivers just can't help speeding up when it comes on!

The irony is that 'breaking the chain' is one of the great behavioural safety principles. It's also a problem for the safety and health world because it's very difficult to prove a negative.

To work with the above health example only 50% of people who smoke are killed by it – so the question is begged if you give up after many years, then avoid cancer, heart disease and lung disease and live to a ripe old age. Was it giving up that did it? Or were you going to anyway?

In safety, many of us have been saved by a hard hat as something has fallen on our head. It's a hell of a shock on the day – I had a stiff neck for a while and needed a new hat – but we survive and we know for certain what broke the chain. But what if someone rushing or carrying something and unable to see their feet kicks a loose brick, power tool or metal clip into a toe-board directly above an area of the site that should be isolated but isn't – and we're stood below with our hard hat in our hands enjoying the sunshine?

They might look over and muse 'whoops, that might have been close!' but we won't even know this happened. (And as fate would have it without that toe-board we'd have died instantly.)

That chain principle applies to a million and one other situations – from driving (if only she'd checked the tyres/not answered that phone/not been so close to the car in front/not turned to shout at the kids …) to getting home late at night (if only they'd crossed to the well-lit side of the street/not gotten into an unlicenced taxi/ not been quite so drunk ….)

You get the idea.

Abba sang a great song about 'gods above deciding' and 'us down here abiding' but sod that. It does, I'll accept, apply overall to luck *generally*.

Because, plain old-fashioned 'wrong place wrong time' bad luck does of course exist. (Like the people killed a few years ago by a falling helicopter walking to work in central London.)

So, whilst we get no guarantees either way: We, as *individuals*, do get lots of control over how much luck we might need.

# No guarantees either way – but we, as *individuals*, do get lots of control over how much luck we might need.

Another home example.

The first sunny weekend in spring parents rush out and buy trampolines for their children ... and the hospitals *know* they'll see lots of very seriously hurt kids that weekend – many with life changing injuries. (It could be failing to set the safety sides up correctly – or at all. It could be faulty equipment, it could be lack of supervision coupled with risky horseplay ('hey guys, I've just thought of this great (and exciting) game') ... it could, as above, just be sheer bad luck.)

This is the key thing: the number of extra trauma doctors and nurses needed they can predict ... the names of the devastated patients and children they can't.

The number of extra trauma doctors and nurses needed they can predict … the names of the devastated parents and children they can't.

**An Attempt at Some 'Mystic Meg' Accident Prediction**

I'd like to try and predict the next accident you, your family or a colleague is involved with.

You see, people tend *not* to have accidents when they are doing something dangerous. Most usually, they are focused and they follow the safe system of work when the risk is substantial. Instead, they have accidents when they are doing something only ***moderately dangerous*** – especially when they have habituated to the risk. (I hate the term complacency. For me that's far too judgemental – and usually said with hindsight.) Over time, we just get more *comfortable* with the risk at hand and as we do our focus can drift.

And … when doing this moderately dangerous thing we may be *rushing* – and/or maybe *angry*, *tired* or *distracted* … or even all of them …

… so, we might well drift into the line of fire or lose control of the equipment we're operating – or just our feet. (See falls above.)

And cue the rolling dice and then … the blue lights.

Driving, as briefly mentioned above and in a later chapter, is the most obvious day-to-day example. (Pilots hardly ever cock up and crash on take-off and landing … but tired, they crash their cars on the way home from the airport.)

Just please recall that moderate risk tends not to be fenced off and locked away. Dealing with it is often left to our own focus – and especially our habits. So, one of the most important bits of advice in this book.

Develop good habits about moderate risks!

One of the most
important bits
of advice in
this book …
develop good
habits about
moderate risks!
They tend to not
be fenced off or
locked away …

**Finally – A Positive Case Study**

Arnie Schwarzenegger has done ever so well for himself. He was first a world champion body builder, then the world's highest paid movie star for a decade *and* finally the political leader of the world's 6th largest economy (California).

In his global bestselling book about how to be successful (*Seven Tools for Life*), Chapter 3 is called "**Work Your Ass Off**"!

You can't argue with the Terminator! Or with Dirty Harry for that matter who once said to an adversary:

*' ... in all the excitement I've lost count ... have I used all six bullets or only five? ... so ... do you feel lucky ... punk? '*

The message of this chapter is ... bugger that. Simply don't fall out with the Dirty Harry's of this world!

And leave the 'punk' stuff to Johnny Rotten ...

# 2 *Why* We All Take Risks

In late 2024 I read in my daily newspaper a reasoned and articulate response to recent government inheritance tax changes by a billionaire businessman (not unrelated – his sons are lined up to inherit his business)!

When I worked with his company he asked me a great question at the end of the session with his board.

> 'He challenged "this is all really good stuff Tim ... but what about the f\*\*\*wits?"'

My slightly risky response was 'well it must be pointed out that you're a f\*\*\*wit yourself Andy, so are your sons here ... so am I for certain ... and so is everyone else in the room too ... the truth is everyone on the planet is capable of f\*\*\*wittery! ...

(Followed by 'please let me explain ....' *and quickly at that!*)

This chapter is about why we do things we really shouldn't do:

## 3.1 ABC Analysis point 1 – or 'why we really don't like to faff about'.

> *'I can resist everything except temptation'. (Oscar Wilde)*

> *'What I like to do with temptation is give in straight away. It saves on the faffing about'. (Stephen Fry)*

ABC (or consequence) analysis shows how we're all capable of giving in to short term temptation and coming up with a short cut or work-around if the safe or healthy way is slow, inconvenient or uncomfortable.

Some more than others, of course, but generally although we usually get away with it, we are *all* tempted to roll the dice from time to time.

DOI: 10.1201/9781003594420-2

# 'What I like to do with tempta-tion is give in straight away. It saves on the faffing about'. (Stephen Fry)

But, as described in the last chapter, we *don't* always get away with it, however. Then, often, we look at others when things go wrong, use a bit of post event hindsight and think 'well, when you think about it, the risk was obvious and the consequence close to *inevitable'*.

In doing so, we can totally ignore the fact that we ourselves take similar risks but have been luckier – or perhaps, more accurately stated, less unlucky.

PPE is a perfect illustration. You may be surprised to know that in the UK for every person we kill in an accident some 100 will (go on) to die a premature death because of something they were exposed to at work. (And that is considered a *conservative* estimate.)

This is how that happens.

Most organisations struggle to afford the very best PPE and will by nature oft only provide merely 'legally adequate' PPE. This works well if:

- The worker pays full attention at fit training so they know exactly what to do (and ideally *why*).
- They always do what they've been trained to do when exposure is possible.

And

- They change the mask/filter as frequently as the manufacturer recommends.

And even that assumes that both the risk assessments and laws are correct. I mean what could – and does – go wrong? (The figures above illustrate with crystal clarity just how often.)

So, if you work near a sign or document saying 'PPE required' (especially respiratory PPE) remember what the desk Sargeant of Hill St Blues said:

'Be careful out there'.

**An illustrative ABC Exercise:** I like to read out a list of risky everyday behaviours at conferences. Behaviours that lead to *tens of millions* of deaths worldwide annually – namely from diabetes, cancer, heart disease, HIV, liver cirrhosis and car accidents. (The behaviours: driving too fast, smoking, drinking, taking illicit drugs and unsafe sex.)

All audiences will admit to a decent score, on average, and will laugh out loud (some with guilty pleasure and some with clear relish and *pride*) when I ask if anyone was tempted to shout 'house!' (The bingo call meaning *'all of them'*.)

You see, we think triggers (antecedents in ABC terms) determine the behaviour (B) but far more often it's the consequences (C). Consequences can be

- Soon or delayed.
- Certain or uncertain … and either.
- Positive or negative.

Anything with a soon, certain and positive consequence will most certainly tempt you. And where there's a temptation there's a head count pointing straight at the chapter one triangle above!

Consequently, if the safe or healthy behaviour required is **slow, inconvenient** or **uncomfortable** in any way, people will be tempted to cut a corner and/or come up with a workaround. From there it's just a head count. It's a short sentence about a simple truth … but it kills millions every year.

Doubling back to that 'pride in bad behaviour' reference leads us to point two of my answer to Andy …

### 3.2 (Because) Point two – being a f\*\*\*wit (sometimes) is *species adaptive*.

If you're reading this, you are a member of the most successful species in the entire universe. One of the reason's we've done so well is that we have a built-in degree of risk tolerance. (Einstein, slightly paraphrased! 'a ship is safest in the harbour but that's not why we built the *bloody* thing!')

More than that, we have a built-in risk *appetite* that we're *born with*. For example, multi-million selling writers like Dan Peterson have pointed out that if small children are provided with a playground that's too safe, they'll adapt it by climbing *over it* rather than through it.

Why? Because it's often fun and exciting to take risks, of course, but it's also reassuring on an existential level as we demonstrate mastery of some of the many risks we face in life.

Some have more risk *appetite* than others but generally, as a species, we quite like a bit of risk.

So, saying or being told 'don't do that/watch out for that … it's risky' might not get you very far at all.

*'A ship is safest in a harbour but that's not why we built the bloody thing!'*

*(Einstein(ish))*

### 3.3 Point three – we're all f\*\*\*wits (5 minutes an hour or more) by physiology.

Juan Fangio could have been involved in the worst accident in motor racing history (88 killed at Le Man in 1958). But despite coming around a blind corner he slowed. When asked how he knew to slow explained that he was Juan Fangio, five times a world champion in the last 7 years, driving a bright silver Mercedes at 200 mph and noticed no one was looking at him – so he knew *something* was up.

That pretty switched on, isn't it? … but, again, we can't simply say 'be as switched on as Fangio at all times' – and this is as good as it gets.

This pie diagrams below show how well we're able to concentrate, on average, on a really good day. (Everyday example: you'll notice sports events such as football, rugby and hockey have a half time at 45 or 40 minutes. The professional athletes involved can run for 80 or 90 minutes no problem – it's the referees and crowd that need a break!)

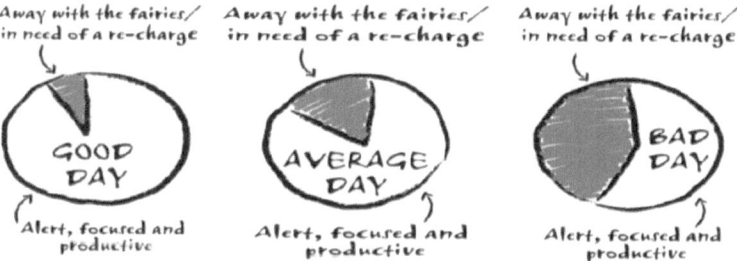

So, even someone having a really good day will spend at least an average of five minutes an hour distracted, away with the fairies and therefore vulnerable to error. (That's someone who has slept well, has no relationship, health or money worries. Who isn't menopausal. Who isn't struggling with their mental health and who isn't suffering chronic pain or on medication. Who enjoys their job, has the tools and time to do it well and who gets on with their boss and colleagues.)

However, if we're suffering from *any* of the items on the list then being distracted for only 5–10 minutes a day can feel impossible.

And we've not even touched on 'long-COVID' – just officially classed as a disability – and its impact on mental health, brain fog and fatigue …

### What to Do about This Physiological Truth?

At the most basic behavioural safety level don't just exhort staff to pay full attention at all times and stay alert to such as trip hazards!

Better to say:

> *'in the 55 (30!?) minutes when you're switched on and alert, please clear up any trip hazards you spot so they're not there to trip over 20 minutes from now when you turn a corner distracted...'.*

More generally, we need to accept that (especially) in a fast paced, changing and dynamic world human error is 'baked in'. We need to pro-actively identify where it's most likely to occur and work our 'Bow Ties' (see above) up the safety hierarchy to identify design solutions pro-actively not just reactively when something goes tragically wrong.

### Three Sports Stories about Error to Illustrate

**The Rider.** I shared a stage once with an ex professional jockey who introduced himself as the 'losingist' jockey in all riding history. (He's Irish.) A man with no less than three world records he announced. The first is for losing horse races – a record 14,000+ races where he lost. (Often, he was riding the favourite too!) The second record is for falling off horses!

He added that he'd pretty much broken every bone in his body …

So, you'd not be too thrilled to have him ride your horse with an important bet on it … maybe the last in a rolling five race accumulator where the first four have won.

Well, until he tells you what his third world record is:

You see, his name is Tony McCoy and he finishes his talk with 'mind you I'm also the *winningest* jockey'. His third world record is his 4,200+ winners.

It's a truth that the harder you try to excel the more likely you are to slip and fall off … literally a horse in this case!

**The Tennis Greats.** Perhaps the three best tennis players in history are Federer, Nadal and Djokovic. Not much doubt about that … but, the percentage of points they won in their entire career?

54.1; 54.4 and 54.4 respectively.

In other words, rounding up, all three lost essentially every other point they played.

Double (faults) all around …

A great book about this is *Black Box Thinking* (Matthew Syed) who illustrates how organisations and societies flourish best when they have a really adult, analytical and objective approach to error. (Another of his books, *Bounce*, covers the content of Chapter 1 above, on the nature of luck, really well).

This approach points companies at a genuinely 'just' and fair culture but it applies everywhere. A multi-million selling book about maximising individual, family and business potential (*Mindset* by a Carol Dweck) says we need three questions to carry with us everywhere:

*   What is going/might well go wrong?
*   Why?
*   What are we going to do about that?

So, this little section has two aims. One to exhort management to minimise knee jerk blame and, as above, embrace what is called a 'just and fair culture'. And two, at anyone seeking to maximise their potential/the potential of their colleagues and loved ones.

(Even if you're the most junior member of a small team, you're still the world expert on what you do ... and by far best placed to answer Carol's 3 questions).

### A Positive Bow Tie/Human Error Risk Management Case Study

The weekend that Ayrton Senna and Roland Ratzenberger were killed the Formula One powers that be said 'this can't happen again, it'll kill our sport'. (It feels that everyone says something like that following a major event, but with billions of their own money at stake (perhaps) they really meant it!) This followed by 'we have the best motoring engineers in the world – they are now going to apply their skills to safety as well as to speed and then share with other teams'.

That was in May 1994. That July, the famous Benneton pit lane fireball happened and one of the main causes proved to be a junior mechanic had removed a filter to speed up petrol flow by one-tenth of a second. He thought he was being helpful ... and Formular One insiders say that recognising the need to involve and empower the front-line engineers proved just as important as the leadership piece.

In short, they came up with a classic 'top-down' and 'bottom-up' culture improvement piece.

The result: in the 31 years since only one person has been killed – compared to an average of 1.2 a year for decades. That's some 'step change' isn't it?

The key point is that not a single driver has slowed down one iota, they're just, as an industry, managing the risks better.

In short, we do not get anywhere saying 'stop being such a f***wit!. Instead mitigate the inevitable effects of f***wittery by involving, empowering, learning and designing pro-actively. We *facilitate* human error reduction with analysis, thought and actions.

So, ... we put fences around trampolines rather than tell kids to make sure they don't fall off ...

As above, at a cultural level we need to involve the real experts (the front line) as much as possible in this analysis and solution generation – an approach that also generates improvements in empowerment and engagement. (Setting the culture for this generally we should also train all management in emotional intelligence skills like assertion and active listening – and on the importance of praise and how to give negative feedback constructively. But that's another book!)

In short, we need to create an environment and culture where our natural propensity to f***wittery is less likely to occur and less likely to cause loss and hurt when it (inevitably) does.

Formula One —
the key point is
that not a sin-
gle driver has
slowed down
one iota since
the Senna crash,
they're just,
as an industry,
managing the
risks better.

**A final point about f\*\*\*wits (aimed mostly at management).**
In these days of staff shortages – especially *skilled* staff shortages – the last point worth making about f\*\*\*wits is that getting the cultural and design issues outlined above right means that organisations are less likely to have to ***knowingly employ one!***

# 3   Culture and the Power of Environment …

*'What do you mean no man is an island?! I am … I'm bloody Ibiza' protested Hugh Grant in the film* About a Boy – *but he turned out to be wrong.*

### A Truth about Culture

In Chapter 2 above I started to talk about an organisation's culture. They don't say 'culture is king' for no reason and I need to address that a little here.

It's not what we say it is, it's what it *actually is* with, for example, new starts and subbies taking a bead on anyone with confidence and/or experience and/or charisma. No matter what the induction says – they'll be highly likely to quickly copy 'what's typical around here'. This means it's day to day behaviour of colleagues that sets the tone – and that means everyone contributes all day every day whether they want to or not.

This, of course, means you too.

You're only 18 and it's your *very first day*? Well OK. But are you part of a new intake of 18 year olds and you're one of the more confident ones.

Then this means you too.

Two truths. One: research shows that about 90% of what you get wrong will be caused by the environment around you. Two: whether you like it or not, you are contributing to your colleagues' 90%. Or your peer groups. Or your families.

### To Address Culture

As above, it's often said that 'culture is king' and it is so worth addressing directly for a while. I'm hoping you think that the positive stuff that's described in previous chapters sounds like the building blocks of a strong culture. You wouldn't mind working there? Maya Angelou's famous quote above might event apply …

> *Here they are … 'not about surviving (but) thriving with passion, compassion, humour and style'.*

To be clear, weak cultures are caused by unfairness, poor communication and low levels of objective learning, trust, engagement and empowerment. In such places you tend to see:

DOI: 10.1201/9781003594420-3

- More turnover of the best (IE most mobile) staff.
- More absenteeism.
- More presenteeism.
- Less discretionary effort.
- More stress.
- More accidents and incidents of all types.
- More spurious (dodgy) claims (and, combined with the higher number of incidents, higher insurance premiums).

Whether you're the head of safety or the very definition of a hard-nosed Chief Financial Officer you really don't want that.

All this is the very definition of the 'win–win' discussion. Talking about *wins* …

Strong cultures are caused by fairness, good two-way communication and high levels of objective learning, trust, engagement and empowerment. So, in such places you tend to see:

- Less turnover of the best (i.e. most mobile) staff.
- Less absenteeism.
- Less presenteeism.
- Fewer spurious (dodgy) claims.
- More discretionary effort.
- Less stress.
- Fewer accidents and incidents of all types.
- And, therefore, lower insurance premiums.

Apart from these lists why bother?

(There is a strong chance you have just thought of the 'what have the Romans ever done for us' scene in *Life of Brian*.)

*'Culture eats strategy, tactics and training for breakfast'.*
*(Peter Drucker said that)*

# 'We all con- tribute to the culture all day every day whether we want to or not'.

*(I say that* 😊 *)*

A *very* **negative mindset and culture.** Later on, in the final chapter on mental health, we'll discuss our award-winning 'F I' scale tool box talk which relates to fatalism and indifference. Here I'd like to address a *really* negative mind-set. The FU mind-set.

The FU scale relates, as you can well imagine, to anger and indignation and people will always be sure to let you know what it is – but very usually *indirectly*. (See the above lists.)

To describe an experiment that illustrates this whole concept:

### How to Create an 'FU' Mindset

In an American university study, they gave students $100 dollars and said they had to share it with a fellow student. The rules were – they got to make one offer to split the money. *But one only*. If it was rejected, that was it. Neither got a cent. They found that a 50:50 offer went down a storm 'let's hit the bar!' 60:40 and 70:30 offers were less well received but accepted. (Maybe *separate* bars!)

However, when it got to around 80:20 an interesting thing happened – many people offered $20 said ... 'no'.

Of course, the person making the offer would oft say 'really, I know I'm keeping the lion's share but you've just turned down 20 free dollars'! However, the response: 'Actually, I like to think I've just *spent* $20 on the unutterable pleasure of being able to tell you, you miserable, selfish, tight assed so and so to ... (well, let's just say 'do one').

It's an understandable and human emotion – but it's really not at all good for your wellbeing and mental health if you feel like that *every* day.

And it's really not good for any sort of culture or the long-term viability of the company either ... you do not want to be working for a company with lots of people with high FU scores. You might well call it a 'toxic' culture.

# 4 Driving

In Chapter 1 on the nature of luck above, we alluded to some key driving behaviours. There's a good reason for that – driving is very probably the most dangerous thing you'll ever do. Statistically, it's hundreds of times more dangerous than flying. (And even if you don't drive – you are probably often driven by a family member, friend or colleague who you can influence.)

> *Case study: the French couple who missed a flight from Brazil in 2009 by minutes only for that flight to crash killing all 228 on board. How lucky were they – except that a couple of days later they crashed their car – killing the wife and leaving the husband with a life altering injury.*

Every year some 1.2 million people around the world will die on the roads. Compare that with the number of people killed by terrorists and the cost and effort we go to protect ourselves from them. Obviously, there's a question 'define terrorism' to answer but by most (western) definitions it's maybe 2,000 people world-wide since 9/11?

Whichever way you might define terrorism though, it's not 30 million plus since the turn of the century. (When you start with actual hard data the way the world responds to risk is often pretty illogical.)

Again, driving is probably the most dangerous thing you'll ever do. So …

**Top Tips for Roads**

• *The* Golden Rule.

The most important rule of them all relates to the chapter above – *don't roll the dice!*

The most dangerous roads are what we in the UK call 'A' roads – where you can travel at real speed even though the only thing keeping you from on-coming traffic is a painted line and where overtaking is often allowed. (So, where a head on crash with a combined speed of circa 200 kph isn't prevented by a barrier.)

And we often *do* need to overtake on these roads to get past trundling lorries, buses and cautious people who are (genuinely) driving just too slowly.

DOI: 10.1201/9781003594420-4

On these roads apply the rule not to ever overtake because you *think* you can safely only when you *know* you can safely.

If that 'think I can do it' turned out to be right - but *only just* and/or because someone oncoming helped you out by breaking or swerving, you'll have been instantly sweating, swearing and having a heart rate double in seconds.

And, as above, a chap called Heinrich suggests you really not have that happen too often …

# On A roads apply the rule not to ever overtake because you *think* you can safely only when you *know* you can safely.

- Tyres.

The experts say that the one accident factor constantly underestimated by the public is tyres. This, whether it's the direct cause or merely one of a number of causes of a crash – see the 'chain' principle above in chapter one. (After all, they're the only thing that contacts you with the road when you think about it.)

Ensuring you always have good tread and the right pressure can save your life … and is a really sensible habit to get into. (On any typical given day driving falls into the 'moderate risk' category we described above.)

So, do you know you currently do have the right tread and pressure? You don't … but are about to set off on a long drive? Taking the kids? … just asking …

- Don't Drive 'Drunk'.

Well, obviously, not drunk, drunk – but not 'drunk' either. Did you know that the physiology of a *tired* person is pretty much identical to that of a drunk person. Easily distracted, impetuous, clumsy … accident prone! (Fatigue is arguably the main cause of the biggest explosion in Europe since WW2 at Buncefield UK.)

Driving when tired can be deadly.

Whilst on the topic though – and just to illustrate the point that keeping out of harm's way is all about awareness and control (and that everything is likely to have an unintended consequence) – did you know you are about eight times as likely to get hurt walking home drunk as driving home drunk … per mile travelled?

Of course, the law courts, fellow drivers and others walking home will thank you for leaving the car behind but the already stressed staff at A&E wont.

If you haven't a 'designated driver' along maybe get a (licenced!) taxi?

- Plan.

Almost everything in life goes better if you plan it and build in leeway … not rush at it, last minute. ('Failing to plan is planning to fail' is the old adage.)

You like to read? You have messages to catch up on? You like a swim or a workout? If that leeway you built in isn't needed you can do this in a local leisure centre/coffee shop/the car if you're early.

And you'll not be stressed and frantic and *whatever it is* you've gone to do with very probably go better!

Without a decent plan we often leave ourselves in a situation where we're highly likely to be driving later on that day when hanging by a thread!

- Two Final 'Golden Rules'.

One, always leave the space to deal with the mistakes of others. (Or get into habits that will help you if others around you make mistakes.) A simple example: if you're waiting to turn across heavy traffic don't turn the wheels in the direction you're headed *until you need to* because if someone behind you has break failure/ is distracted ... they'll shunt you straight into oncoming vehicles.

Of course, a good defensive driving course covers that and much else about driving heads up and to the distance for advance warnings because it's all about antica ....

... *pation* (as Rocky Horror says).

So going on one if you possibly can ... is a bloody good plan!

Two. Related to one ... assume *everyone* is a drunken idiot (because frankly lots are). Apply this rule to pedestrians too!

Always give yourself the time and space to deal not just with your own mistakes but the mistakes of others too – assuming they're a drunken idiot helps.

### *The* Big Red Flag

Speeding has it problems but this book isn't focused on 'follow the rules' it's focused on personal *risk management*. The big red flag is the driver who needs to *brake sharply* ... because that might be to do with excess speed – or it might be inattention/fatigue or it might be not leaving enough space – or all three.

Basically, if you get through break-pads far more quickly than the norm then you're much more likely to join the 1.2 million a year mentioned above. You have the reflexes of a cat and the eye hand co-ordination of Roger Federer? Well good for you, these do help – but sadly for you all the studies suggest reflexes and eye hand co-ordination cover less than 10% of the issue. Speed/distance cover the other 90%. (And can I please refer you back to the above figures about just how many times in his career Federer screwed up.)

### Be a Climate Hero ...

People who do the above well use less petrol and have vehicles which need less maintenance. So, as well as smooth drivers hardly ever crashing – they're good for the planet too.

> *Therefore, give out generous company awards for 'green' driving – if you run a large fleet, even 'very, very generous' awards will save an absolute fortune in maintenance and repairs!*

### The Bald Fat Advanced Driving Instructor – Assertion and Mindset

Lots of other drivers are *indeed* frankly selfish, ignorant morons who we'd all like to shoot off the road with one of those James Bond style headlights that turn into rockets!

But let's, instead, talk about a mind-set that points at 'fellow-travellers'. It helps the stress levels and minimises the chance of a crash.

Imagine a bald fat driver who spends half his time sitting under a tree! Tolerant, patient and wise and who never react with anger when they leave a nice two second *minimum* gap (as in 'only a fool breaks the two second rule' – four in bad weather) – but some cretin notices this gap and undertakes.

The rest of us can be tempted to lose our temper and maintain this (now *half* a second gap) for a while and even Buddha himself said it was OK to react with righteous indignation.

But unless you're actually going to ram them off the road and into a field all you're doing by reacting badly is increasing *your* risk level. Honestly, you achieve *nothing else* at all. They are **not** seeing that you're angry and thinking 'that teaches me, I was very rude and risky just then and I realise I shouldn't have done it and won't ever do it again'.

In truth, I've always noticed that their hand signals typically suggest a more confrontational and less contrite mindset.

**So, What to Do?**

The fat guy referenced above would say 'bad things are inevitable in life, starting with the fact that you and everyone you love will one day die – it's how you deal with these bad things that counts ... so in this really relatively trivial situation, look at that photo of your children you have taped to your dashboard, take control of your rage, remind yourself that the dickhead who just undertook you rudely has to live the life of a dickhead every day ... chill out, and just back the f**k off. It's only delayed you a second'. (I'm paraphrasing the Buddha slightly but I'm sure you get the idea.)

It's a psychological truth that you cannot think and react simultaneously – unless you're a bit psychotic – so remember the 'photo on the dashboard' trick - or just count to 10 – but stay *thinking* not reacting.

# Unless you're *actually* going to ram them off the road and into a field all you're doing is increasing your risk level.

This an excuse for an illustrative old joke.

> *A footballer says to the referee 'I can't believe you didn't give that ref!' followed by 'what happens if I call you a useless bastard?'. The referee says 'if you do that, I'll send you off'. The player thinks a bit and asks 'what happens if I <u>think</u> you're a useless bastard?' The referee says 'it's a free country son ... you can think what you like'. The player walks off, commenting over his shoulder 'OK then ... I <u>think</u> you're a useless bastard'.*

You can think all sorts of things about the idiot driver: Most positively that they may well be rushing somewhere genuinely urgent so it's best not to assume ... or that luck is limited and karma a bitch and, frankly, they're on borrowed time ... or maybe even that you're glad you're less of a selfish di**head than they are.

Just, please, when behind the wheel, stay *thinking!*

# 5    Your Wellbeing

**First – Overcoming "ABC Inertia" and It's Links to Wellbeing**

Here's a simple exercise. Close your eyes and picture someone you'd quite like to meet – in this example because they look *interesting* not because you fancy them! The physiology is different!

Then, picture yourself in 20 years' time. You might be surprised to know that a picture of your brain waves would look pretty much exactly the same as, essentially, 'future you' is a stranger to you. (Pension advisors find contributions go up if they show clients computer aged photos of themselves.)

When you combine this with the instinctive thinking and behaviour described under ABC (temptation) Analysis above it means it's just really hard to think and act long term.

(So, get one of those aged pictures printed out and stuck to the fridge or near that never used exercise bike in the garage. It'll really help overcome the 'delayed, uncertain' daily inertia! Another exercise tip, join a gym and get into the habit of working out when people you find attractive go there – it'll automatically add 25% or so to your effort!)

**How to Commit to Self-Improvement**

Again, it's possible you'll have seen 1001 books about 'marginal gains' and 'atomic habits' that will in the end, after lots of hard work and dedication, deliver the 'new you'. They all relate to chapter one (you tend to get the luck you deserve depending on the effort you put in) and chapter two (we can so easily get short term sidetracked on any given day).

Pretty much any result simply comes from hard work and focus really.

DOI: 10.1201/9781003594420-5

# Remember Arnie Schwarzenegger's Golden Rule for success:

# Work Your Ass Off ...

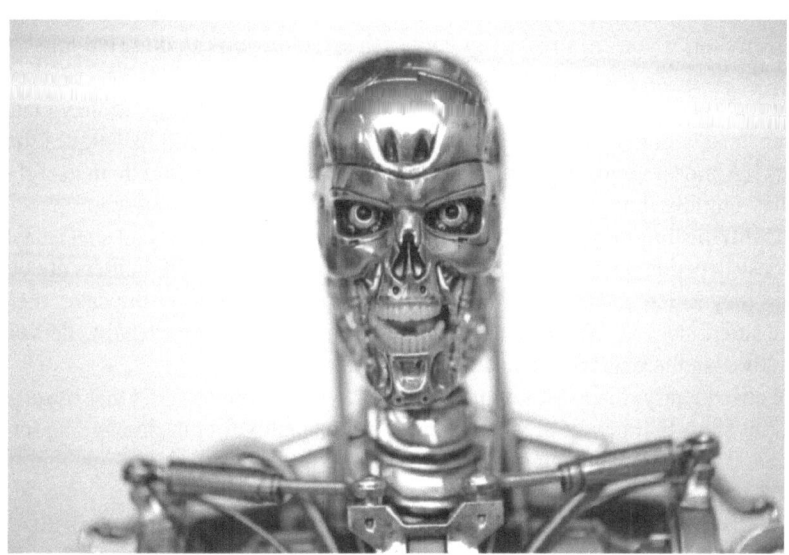

I wonder if the very best advice on what to focus on is this: Pick someone you know (or know of) from the world now or from history. A person you'd really like to emulate and ask yourself what they do/what they would do in a given situation.

Then do that as often as possible.

### The Generic What to Dos

It's highly likely you'll understand all the basics: eating well, not drinking too much alcohol, taking exercise and sleeping well.

Putting these in a basic framework, wellbeing guru Martin Seligman and other experts say that there are five interrelated and overlapping factors that impact on your levels of wellbeing and mental health

* Physical health.
* Family and friends.
* Money.
* Feeling you contribute (generally).
* Having a job that you enjoy.

Some quick notes on the (less obvious) aspects of the above.

The first is that you do, of course, need **enough money** not to be stressed. But its benefits plateau quickly. Having lots of money doesn't necessarily make you happier than having merely *enough* and a remarkably high percentage of lottery winners have found it ruined their lives. (Though yes, I'd be more than happy to risk it too!)

However, if you are stressed by your cash situation you can do two things today. One, re-read Chapter 1. It applies to _everything_ including career progression and financial success. Two, attend (or watch online) one of the many 'money management' talks/workshops. They are stuffed full of useful tips and techniques that *do* work. (Almost everyone who watches/attends them says they find them useful even if they initially thought they wouldn't.)

**Contributing** may seem less obvious but we are a hugely social species whose success depends on a collective effort. Back in cave days if the hunters ate everything they killed and took nothing back for those looking after the cave, then the cave sitters starved. Worse, if the hunt is unsuccessful and the cave dwellers eat all the stores in the hunters' absence then *everyone* starves!

Consequently, the most important social concept of them all is that of *fairness*. Studies show that people are judged much more harshly for unfairness than for illegality. (Laws just being an often clumsy, and biased toward the people that wrote them, way of getting us to act fairly.)

Whilst people can come up with all sorts of mental gymnastics to justify their actions – and some others are just well, *horrible people* – most want to feel they have pulled their weight and done their share.

What to do? You might consider voluntary work (of course). You may need to/ have chosen to help or care for family in your spare time. You might help coach the kids now you're too old to play yourself – or even as well as playing as you end a career.

Discretionary activities at work like joining a safety committee or training to be a mental health first aider tick the box too. Or maybe just doing a good day's work – especially in a job that's large part vocation perhaps. (Nursing? Teaching? Working with animals etc.)

I don't want to sound all new age and fluffy but it's true that it's very easy to be so busy making a living/getting by/getting through that we forget to 'feed our souls'. (Or some such!) But a quote I once read nails it …

*I've just realised I spend half my life earning money to buy things I don't really want or need in order to impress people I don't even like'.*

## Your Job

*Because good work is good for you. It can give you traction, purpose, meaning satisfaction and, if you're lucky, a damn good laugh with like-minded colleagues.*

Here's a simple check question to see if you are in a job that suits your personality and/or your skills. How often are you in 'flow' where you're absorbed in what you're doing and time passes quickly?

'Hardly ever' simply isn't good for your wellbeing.

And, finally, here is a checklist of things we know can cause stress if they're not matched to you. (Over and above the obvious such as unrealistic workload, being indoors when you like to be outdoors – or viz a viz – and/or not having the tools and/or training!)

If you score badly on *any* please consider talking to your manager about rotation or job design – or even changing job if you can

- *Autonomy and control.* You like lots but you have little? You like a little but you have lots? In both cases you're going to be stressed.
- *Skill development/creativity/variety/interpersonal contact.* Not matched to you? Then, you're going to be frustrated …
- *Meaning and purpose.* To what extent do you feel like a cog in a machine that delivers something of little value or meaning to you?
- *Role clarity and uncertainty* … and a cog kept in the dark at that? (Please look up the old joke about the 'mushroom treatment'.)

Note: If you can't impact on much of the above (and are scoring badly) then you *really* need to focus on the stuff you can control like eating well, exercising, sleeping well and meditating etc. etc.

Essentially, you'll really need to consider how to manage your stress and 'bust' that stress when you can. (Hammering the hell out of a punch bag is a lot better for you than hammering the beer and running to bust stress and for fitness a lot better than running away from your problems … you get the idea.)

Though it must be noted that the Scandinavian concept of Hygge was very trendy for a while. And a bloody good thing too as it's based on the notion that a little of what you fancy does you good ... (even alcohol, chocolate and naked saunas etc.).

It's the emphasis on *moderation* that's the tricky bit!

*'I've just real-ised I spend half my life earning money to buy things I don't really want or need in order to impress people I don't even like'.*

### The Overlap with 'Psychological Safety' and Work Culture

Lots of organisations are striving these days to be 'psychologically safe'.

First to address shouts of 'psychological buzzword bullsh*t' and other 'snowflakery'.

It's true you simply can't please all the people all the time and some people do, of course, need a far greater sense of perspective than they seem to have.

However, with almost no one disagreeing that one in five of the work-force are genuinely struggling with their mental health then it's clearly a huge problem when the workplace is 'unsafe'. (Even halving that estimate to 1 in 10 to be more robust in our definition of struggling and we're still talking about 350 million workers worldwide – and the next chapter gives some hard facts about how that plays out at the sharp end.)

It's just true that people who feel psychologically safe at work are more likely to be productive and less likely to be stressed, anxious and distracted. These are all good things, so I'd like to cover the root causes of psychological safety and I hope you'll agree they are all win–win items that generalise to all sorts of issues. (See the lists above.)

But first to address the sharp end. Out and out **bullying**.

To steal from a famous film scene that covers a working definition of bullying:

'why are you upset? we're not laughing at you we're laughing *with* you', 'but I'm not laughing', 'OK, then we're laughing *near you*'.

Leaving it at that is bullying. Apologising (genuinely) if you've upset them then talking to them with empathy to see if it's them being over-sensitive ... or it's you being *insensitive* is being a reasonable adult.

So, of course psychological safety means not being bullied but more day to day it essentially covers three things.

- Being unafraid to pass bad news up. (What you don't want: someone who asks aggressively at the end of a meeting 'anyone got a problem with those cuts/ changes? No?! Good ...!? Get on with it then ...', then blames you for 'not pushing back enough' when it goes belly up.)
- Being treated fairly. Especially when things go wrong. (See lots of the material above.) What you don't want: at the BBC it was said that when a poor *executive* decision led to a cock-up the shout would go out that '*junior* heads must roll'.
- Being unafraid to ask for help when you need it. (What you don't want: 'oh, really? Give me strength. Just what I need to have to deal with today ...'.)

You'll see that that the first and second points link with Chapter 2 and the third point with the notes above in this chapter. Asking for help gives the person you've asked *the chance to help* ... so it can be a double win. (Again, all the stuff in this book is, as above, inter-related and overlapping.)

It's estimated that about 1 in 5 of the workforce are struggling on any given day – that's about 700 million people worldwide.

## How to Have a Positive Mindset

My business partner Jason Anker MBE paralysed himself in a fall from a ladder 30 years ago. He first talked about the accident and it's tragic and far reaching consequences but now gives a wonderful 'inspirational' talk about his current happiness and contentment. It's hung on his "3 As" which he researched in detail and which encompasses all the classic thinking in this field.

The As are:

- Acceptance.
- Attitude.
- Action.

**Acceptance** is essentially about sitting under a tree (or some such!), centring yourself and thinking clearly about how things *really are.* (Again we can consider those vital questions suggested by Carol Dweck. What's going wrong? Why? What are we going to do about that? (Or alternatively – what's going well? why? how do build in leeway so that continues even when things get difficult?)

Then, in the light of that, simply thinking through with clarity what might need doing and why. (You'll note there's no mention of meditation and/or mindfulness and the like here – though these both prove very effective for many people. It's just about being present in the moment and clear in your thinking. A cup of tea and a quiet sit? A long walk in the hills? A soak in a hot bath? An hour's gardening? A hard work out? An hour with a jigsaw? Maybe even a stiff drink or two … if it genuinely works for *you* ….)

**Attitude** means not letting fear hold you back but being positive, being realistically ambitious and most of all being *grateful.* As Jason says maximising thrive in life is 10% what happens but 90% how you react to that.

**Action** relates to taking those daily (oft very small, incremental) actions which will maximise your wellbeing and which are (and often *need to be*) powered by the right mindset. (Of course, again see Chapter 1.)

I think attitude and action are pretty self-evident really (and the practical tips below I hope also help to illustrate the principles) but two quick illustrations of *acceptance* which is a more nuanced concept.

The first is personal to Jason's wheelchair – an object he understandably hated for decades – the first thing he saw every morning and an instant daily reminder of his disability. However, he has come to genuinely **accept** it simply isn't something he can change – and so actively chooses to see his wheelchair as a tool for mobility, travel, independence and productivity. An enabler, a friend – a valued personal assistant if you would. (As a successful speaker he is making money that is allowing his two granddaughters – to whom he is *beyond devoted* and who always kick off his on 'grateful' list every day – opportunities he never had himself.)

The second is the story of the Kintsugi vase. Many centuries ago, an Emperor's favourite vase got broken by some servants and though they tried hard to fix it, the faint cracks still showed on close inspection. Since close inspection might have ended very badly(!) they had, in panic, the idea of highlighting the cracks with gold emboss. Luckily the Emperor *did* like it and soon, all around the kingdom vases were being smashed and repaired. Or so the story goes. Certainly though, in time a beautiful philosophy has grown around this:

- Point one: you might well end up with something more nuanced and beautiful than you started with. But …
- Point two: it will, like each of us, be *entirely unique.*

Though it must be admitted that not many of us are quite as unique as our Jace.

### Some Top Tips for Positivity

*Have SMART Goals for Yourself*

Have a specific, measurable, achievable and realistic goal for yourself. (With a realistic timescale attached.) Knowing you're working towards this will give a great sense of purpose and satisfaction.

(There's a school of thought that says what we think of as a 'lack of willpower' is just an unrealistic plan crossed with not genuinely wanting the thing in question as much as we say we do!)

### *Start Every Day Thinking of a Couple of Things You're* Grateful *For*

Nothing points you in the right direction like a bit of gratitude … (for those of us getting older perhaps just that we are waking up at all!). If you find that difficult – just think of anything or anyone you'd hate to lose – or about anything that could have gone far worse than it actually did (especially if the dice were rolled and you got a bit lucky).

### *Remember You're Probably Deluded and Overconfident (So Make a Bloody Effort Why Don't You?!)*

Studies suggest we all think we're, for example, a 7 or 8 out of 10 when actually we're a 5 or 6 (or worse)!

Not you of course? OK, try these: how good a car driver are you? How good a lover are you? How bitchy and mean spirited is your gossiping? (Studies show only about 5% of readers put themselves in the worst 50%!)

*You don't need to be a top statistician to note that the data doesn't match up!*

In truth, *actual* 8 out of 10s tend to be worrying they're only a 5 as they strive to push on to 9 … which, to be slightly fruity (see one of the questions above) others might appreciate in all sorts of ways!

### *Remember That Behaviour and Attitude Breed Behaviour and Attitude*

*'Smile and the world smiles with you … cry and you cry alone' is the old expression.*

It's a truth that, through self-fulfilling prophecies, people who expect the best of others very often bring out the best in others. (Praise is about 20 times as effective a way to change others behaviour as criticism and coaching about 12 times as effective as telling. You'll see lots of books from successful people talking about raising themselves up on the shoulders of others.)

The bestselling book (*How Spies Think*) has the wonderful (wise) advice:

*'always assume the best of others to try to always set up positive self-fulfilling prophecies … but do always check the important stuff'.*

### *Balance Your Buts*

'This person is usually OK but today they're really p\*\*\*ing me off' will point you automatically at criticism and judgement. 'This person is really p\*\*\*ing me off today … but normally they're OK' will point you at empathy, analysis and compassion. (The important stuff is always after the 'but' … the words before the but usually just waffle and flannel.)

So, make 'balancing your buts' a habit … especially around safety and health. The request 'safely but by Friday please' will usually get the task done by Friday as safely as is viable in that time-scale. (And cue work-arounds, short cuts and rolling dice.)

### Gossip Positively

We all love a gossip. But, get into the habit of at least *starting* and ending a good gossip with something positive and kind. (Or to quote the famous film line 'Oh, for f\*\*k sake could you not just *BE NICE* for once!')

### Use the Two-by-Two Box

Imagine a two-by-two box about the strengths and weaknesses of two sides of an issue.

The 'my strengths' and 'your weaknesses' sections are really easy boxes to fill in – but the 'my weaknesses' and 'your strengths' boxes are far more useful lists. (From preparing for a negotiation, to learning about yourself for personal development.)

### People, Places and Things

A phrase any recovering addict will recognise.

We know (see Chapter 4 above) that 'culture is king' and that about 90% of what people do is driven by the environment around them. Those people you admire and would like to emulate? Go and hang out with them best you can …

'*Always assume the best of others to try to set up positive self-fulfilling prophecies ... but always check the important stuff*'.

*(David Omand, UK's former top spymaster)*

# 6　Mental Health

The very definition of last but certainly not least ...

To re-iterate, in the UK you are 32 times more likely to lose a colleague to suicide than to an industrial accident. It's around 10 to 1 even in high hazard industries like construction.

(The UK scores really well internationally for safety but is mid table in terms of suicide figures. So, though other comparable countries have less severe ratios, it's only because their accident figures are worse.)

## The Mental Health First Aid (MHFA) Controversy

Here's a direct quote from a conference debate I chaired at the NEC Birmingham:

> *'We have MHFAs at my company and senior management seem to think that's got it all covered. But the thing of it is – they still carry on treating us like crap and we're all really stressed'.*

An illustrative anecdote: many decades ago, I worked briefly at Llanwern Steelworks in South Wales. I actually went there on a moped but please imagine my mum dropped me off on day ome to be told:

> *'It's true Mrs Marsh that we have hot metal splashing about, noxious gases all over, fork lifts speeding around like it's a Grand Prix and a horrible bullying management culture – but young Tim will be safe with us as we have no less than four highly skilled first aiders out there ...'.*

I'm not sure *she'd* have been all that reassured ... and I'm sure *you* get the point I'm trying to make.

MHFAs can help – especially if they are an indication management take employees mental health seriously – but they should be seen as but a small cog in a holistic and integrated approach.

## What *You* Can Do

For *yourself?*

Well, see Chapter 4 especially (though elements of all the others for that matter).

DOI: 10.1201/9781003594420-6

## For Your Colleagues and Friends

### *Be Aware*

Many people spend more time with colleagues than with family and friends. So, the first person with the chance to spot a warning sign will probably be a peer at work. Has a chatty one gone quiet? Has a quiet one gone chatty? Has a tidy one gone scruffy? Does someone repeatedly joke about their mental health and/or suicide? (Returning to the same subject, even in a light-hearted way, is an indication that it's on their mind.)

Just be mindful of that one in five figure and keep an eye out.

You have 20 colleagues you see regularly? Remind yourself you probably need to be worrying about four of them …

### *Ask Twice*

There's a very good chance that the first time you ask 'are you OK?' of someone (even someone struggling badly) they will respond with 'I'm fine thank you … how are you?' (In certain countries and age groups responding like that is close to a *law*!)

So, **ask twice**. Maybe straight away with a 'are you *really* OK? As actually, you look pretty distracted'. Or perhaps by asking at a better time and/or in a better location.

Ninety-nine times out of 100 you can safely accept that 'fine' answer and crack on even though you can tell full well that they're *not* OK. You can reassure yourself that they most probably will be OK (and you'll probably be really busy).

But maybe 1 time in 100 you'll carry guilt to your grave.

### *Ask a Gateway Question*

Asking, for example, 'how did you sleep last night?' often produces the response 'totally crap actually!'

You can then ask *why that was* …

Other gateway questions might be such as 'anything on your mind recently?' or, if you know the person and their interests try 'looking forward to Saturday?' If the answer is 'yeah … suppose' when it should be 'you joking?! Can't wait …'.

### *But Do* Say Something

We often feel anxious that we'll 'get it wrong' or 'make things worse' but research shows that it's almost impossible to do that.

Talking someone down from a ledge has about a 50:50 chance of success even with no training. Jon Bon Jovi famously did it in 2024 – and let's skip the joke that he threatened to start singing unless they stepped back! (But it's well worth noting that this 50:50 is several times better than the chances of saving a heart attack victim with a defibrillator outside of an A&E unit and we'd always give that a go.)

So, get stuck in and say something ... *anything*.

### Ask About Their FI Score (Better Still Run Toolbox Talks on the Topic)

The FU score we covered in the chapter on culture. ...Here, I'd to tell you about our award-winning "FI" toolbox talk. You see it's really hard to talk to someone with a high FU score but you *can* usually get someone with a high FI score to talk to you. The trouble is, even so, a high FI score can kill you.

As discussed in the chapter on wellbeing, my business partner, Jason Anker MBE, paralysed himself in a fall 30 years ago. He won his MBE for a talk that's key element was 'if something doesn't feel safe then step back and take five seconds'. His problem is that, on the day he *did* take five. When we discussed it years later, he said that, in truth, the last thought before starting to climb the unfooted ladder was 'oh, f*** it'.

A simple expression that we all use that's usually some combination of fatigue, distraction, fatalism, irritation, impatience ... and lots of other stuff.

We turned Jason's honest response into a joke slide: 'the fatalism and intolerance of individual, societal and organisational stressors scale' or 'FI score' for short ...

... but, to our surprise, as well as laughing when they got the joke clients spontaneously went off and used it as the basis of a toolbox talk.

It works really well and wins awards because (*we realised later*) it:

- Uses the actual language in your head when you're stressed. (Nobody ever says to themselves 'oh bother I'm having a big blue pie day today ...'.)
- It uses humour. (Shared humour is two people 'getting it' together and, the thing is, someone getting you when you're really down can make all the difference in the world.)

Full and complete instructions for using the FI Toolbox talk:

- Explain in broad terms what it is.
- Ask if anyone has a high score on a 1 to 10 scale.
- Make sure to talk one to one with anyone who shouts "12!" before they go and do anything dangerous or important. (Or anyone who stays quiet but who pops their head around your door a little later as invited.)

And a final mental health tip related to the mistake we nearly all instinctively make ...

**Use Kind and Constructive Questions and Listening Only**

Things that work:

- Listening.
- Kind/ constructive questions as above and like 'would you like me to call someone?', 'could I make you a cup of tea' or 'how can I help?'

Thing that research shows work far less well:

- Reassuring them that it'll be OK. (They're almost certainly being deafened by the internal 'no it won't'.)
- Backing this assertion up with your own experience/the experience of someone you know. (As above their internal voice if shouting 'but you're not me'.)

If you *do* have personal experience, and almost all of us do directly or indirectly, just use that to power your ***empathy*** and ***listening***. We call this CERR – curious empathy, then reflect or (if listening, question-based coaching isn't sufficient then *refer*).

At the risk of finishing this booklet with something that sounds a bit 'tree huggy'. In recent years the world seems to be increasingly mad … increasingly *bad* even. Frankly, we need all the empathy, kindness and listening ears we can get our hands on.

The good news is that, post covid especially, more and more people are comfortable admitting that. I hope the above notes summarise how we can all best contribute to that sea change in societal culture.

Because in the end, and to quote the Beatles, it is a law of the universe that the kindness we get collectively is exactly equal to the kindness we give collectively. So, a very final reference back to the universal lesson of Chapter 1. … Please do be kind to others and be kind to yourself. In thoughts and in deeds.

## Conclusion

People are very complex, nuanced, changeable, contrary, intelligent, stupid and downright unpredictable.

But where they're likely to go wrong and why is actually quite easy to predict. What to do about it hardly rocket science.

The overall numbers we can do quite accurately but the exact names and places … that's much trickier.

One thing that *is* true though: follow the advice in this simple book and you're very much less likely to add to those figures yourself.

So please *thrive* personally. Even better – help facilitate those around you to thrive too.

Finally, thank you for reading! Can I wish you the very best of good luck personally – though please do try and make sure you don't need too much of it!

# Further Reading

If you think the concepts and ideas in this booklet of use and worth delving into in more detail, then I'd like to recommend some books and authors.

Of my own books *Talking Health and Safety* is the (next) most practical and user friendly. Of perhaps the two greats in the safety field I most enjoyed Sidney Dekker's *Field Guide to Human Error* and James Reason's *The Human Contribution*. (Though anything by either of course.)

More mainstream books I'd recommend … Matthew Syed's *Bounce* (covering the luck principle) and *Black Box Thinking* (on the importance of learning) as a starter for ten. Or *Atomic Habits* by James Clear which is a very user friendly book on maximising potential as is Arnie's book *Be Useful – 7 Tools for Life*. Or Bruce Hood's *Science of Happiness* … basically there are lots of good books out there … Just pick the one that sounds the most interesting to you.

Finally, I must acknowledge Gert Gigerenzer. His excellent *Risk Savvy* was, best I know, the first use of the word savvy in this field.

DOI: 10.1201/9781003594420-7